In-kind contributions

The Codex Alimentarius Commission thanks the following Member countries for their generosity as hosts of Codex committees and events.

- Australia
- Canada
- China
- Ecuador
- Germany
- Hungary
- India
- Kazakhstan
- Kenya
- Malaysia
- The Netherlands
- Paraguay
- Republic of Korea
- Senegal
- Uganda
- United States of America
- Vanuatu

Acknowledgements

This publication, CODEX, has been prepared by the Secretariat of the Codex Alimentarius Commission, part of the Joint FAO/WHO Food Standards Programme.

Now in its second year this Annual Report presents the work of the organization in an accessible, colourful and readable format for the widest possible audience.

The Codex Secretariat gratefully acknowledges the contributions and support made by FAO and WHO; Officers of the Codex Alimentarius Commission; Host Governments; Members and Observers and the Codex Secretariat.

Thanks also to the communications and technical units in FAO and WHO for their advice and encouragement and to photographers and technicians for their contributions throughout the year at Codex events.

Editorial team: *Anne Beutling, Giuseppe Di Chiera, Ross Halbert, David Massey and Mia Rowan*
Visual Communication: *Cristiana Giovannini*

A world full of standards

Contents

4 What is Codex Alimentarius?

6 A forward-looking agenda

8 The Chair's10 goals

10 One year in

12 Behind the WHO Five Keys to Safer Food

14 On the 'future proof' track

16 An insider's story

18 When do you need INFOSAN?

20 Nuclear Codex

22 Food fit for the future

25 Part of the system

26 Trade and food standards

29 Building Codex globally

30 20 years of risk analysis

34 Observing Codex

36 Moving on

38 When #Codex trends on Twitter

42 Codex is people

44 Improving cooperation with standards

48 List of Members

50 How the Codex Alimentarius is funded

51 List of standards proposed for adoption at CAC41

What is Codex Alimentarius?

The Codex Alimentarius, or "Food Code", is a collection of standards, guidelines and codes of practice that governments may opt to use to ensure food safety, quality and fair trade. When the standards are followed, consumers can trust the safety and quality of the products they buy and importers can trust that the food they ordered will meet their specifications.

The standards are adopted by the Codex Alimentarius Commission, which currently comprises 188 Member Countries and 1 Member Organization (EU) and 226 Observers of which 56 are intergovernmental organizations, 154 non-governmental organizations and 16 United Nations agencies. The Commission, also known as CAC, was established in 1963 by the Food and Agriculture Organization of the United Nations (FAO) and the World Health Organization (WHO) to protect consumer health and promote fair practices in the food trade.

FAO/WHO Regional Coordinating Committees

Codex has developed hundreds of standards, guidelines and codes as well as defined thousands of permitted levels of additives, contaminants and chemical residues in food. Above all, its success lies in building consensus, working together and making decisions based on science.

Codex mandate

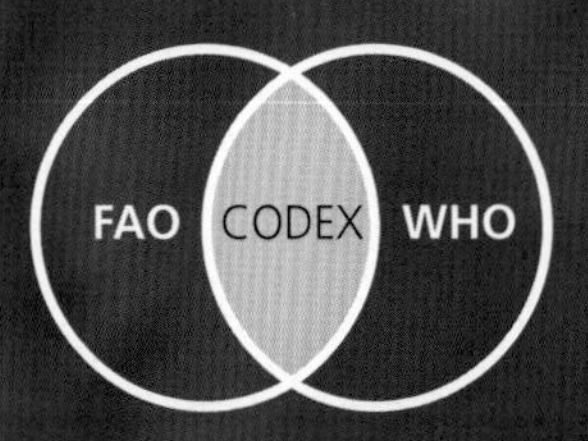

PROTECT the health of consumers

ENSURE fair practices in the food trade

PROMOTE coordination of all food standards

A forward-looking agenda

by Tom Heilandt, Codex Secretary

Great thinkers in traditions the world over, from Confucius to Ovid—not only profound philosophers of life, but also cartographers of the culinary arts and gastronomical sciences—repeatedly returned to a common conclusion: life without friendship and food is no life at all. Wise words – universally true and right in our mandate.

The United Nations deals with food security in Rome! FAO, WFP and IFAD are here and so is the Codex Secretariat.

Let me extend a warm welcome home to FAO headquarters and to Rome for CAC41 – the annual gathering of the Codex family, where we come together to take stock of our collective progress, to plan for the future, and to strengthen our bonds of cooperation and friendship.

Such gatherings not only nourish exchange and discussion, but also remind us of the material links between sustainable consumption and harmonious coexistence, celebrated whenever we eat, be it in a company cafeteria or a street food stall, or at home – everyday scenes illustrating the universal concerns Codex was founded to address, promote and protect.

Our hybrid mandate makes Codex so successful.

©Bob Scott

Codex continuously strives to better respond to the needs of its stakeholders – chief among them, you, Member State partners but ultimately everybody. Without your expertise and material contributions, Codex would not be Codex.

The importance of regional perspectives for multilateral endeavours to succeed at the global level is increasingly recognized throughout the United Nations system, including by Secretary-General António Guterres, whose reform agenda emphasizes the need for decentralized, country-led processes that produce applicable solutions for real-world concerns. Codex – as a hybrid technical Commission privileging country-led participation in setting and implementing its agenda – has long been a trailblazer in this regard.

Our ambitious programme of regional workshops over the past year has sought to provide tailor-made training in our suite of innovative ICT platforms to enhance participation at all levels and continue to improve the inclusive credential, legitimacy and credibility of Codex work. The collaboration between countries, regions and the Secretariat has been exemplary, and I look forward to these initiatives bearing fruit over the year to come as the revitalization of the FAO/WHO Regional Coordinating Committees continues to build momentum.

The past year marked some major milestones for Codex – including the 50th sessions of the Committees on Food Additives and on Pesticide Residues – as well as some significant challenges. Retirement of a longtime staff member and an extended vacancy brought some difficult moments but those are now resolved and the Secretariat is fully staffed. More fundamental issues remain on our agenda: What do we do when consensus is elusive for reasons beyond our mandate? How do we ensure the sustainable availability of scientific advice? How do we ensure that all who wish to participate can equally do so? How to we ensure that Codex standards – once agreed – are actually used.

Many activities are underway supported by FAO and WHO to solve the issues. CTF2 is now in full swing but we need to ensure its continued finances. Initiatives for supporting scientific advice are moving forward but we cannot be sure yet if they will achieve what we hope.

How Codex approaches such fundamental questions, while continuing to produce global public goods so crucial to many countries, is ultimately for you, the membership, to determine. Vigorous good-faith engagement and robust commitment will be key to finding solutions and setting the kind of forward-looking agenda required to enhance our legacy and contribute to the solutions to the existential challenges facing humankind over the 50 years to come. The Strategic Plan 2020-2025 is only the beginning.

Safe food and fair practices in the food trade will remain important forever. To remind us of this we are supporting together with FAO and WHO and many members led by Costa Rica, the creation of a World Food Safety Day. The proposal will go to the UN this autumn and we look forward to making this day a success with all of you.

The tasks before us may seem daunting, but I am confident in our proven bonds of friendship and cooperation to deliver consensus and meet the expectations entrusted to us.

The Chair's 10 goals

by Guilherme da Costa Junior

The Codex Alimentarius is the preeminent international organization dealing with food safety and fair practices in the food trade. Its food standards, guidelines and codes of practice contribute to international food trade safety, quality and fairness.

Furthermore, they are a robust way of preserving consumer health and preventing foodborne illnesses. As global food trade increases, Codex food standards are accepted worldwide and serve as the benchmark standards for the World Trade Organization. When countries base their national standards on Codex, whether for domestic consumption or export, they give consumers confidence about the safety, quality and authenticity of what they eat.

Guilherme da Costa Junior is Chairperson of the Codex Alimentarius Commission

Food safety and trade are in a state of constant evolution. As we look to the future, I urge everyone to stay focussed on the following "10 Goals for Codex":

1. Be ready to respond to emerging food safety, quality and nutrition issues in a timely manner to protect consumer health and ensure fair practices in the food trade.

2. Ensure Codex standards are based on sound scientific evidence.

3. Connect with the realities of different countries, bring countries together and build consensus at every step of the Codex standard-setting process.

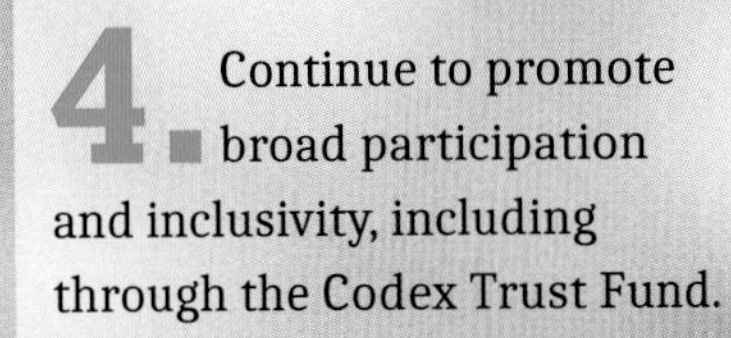

4. Continue to promote broad participation and inclusivity, including through the Codex Trust Fund.

5. Raise consumer awareness and build trust in Codex standards, acknowledging that consumer associations can help our stakeholders see issues from a public perspective.

6. Enhance the participation in Codex work of diverse actors along the food chain to facilitate access to healthy, nutritious food, and provide standards to guide people who depend directly on agriculture and the food system for their livelihoods.

7. Strengthen cooperation with other international organizations related to food safety and fair practices in the food trade.

10. Act together in Codex – in our changing world of ever-swifter communications and scientific developments with significant implications for food safety and trade, Codex works to strengthen collaboration across different sectors, both public and private, through genuine and effective partnerships.

8. Step up advocacy and outreach to all sectors involved in food safety and fair practices in the food trade.

9. Contribute to the achievement of Sustainable Development Goals.

One year in

Messages from the three Vice Chairs following their first year in office

Mariam Eid

To be a Codex Vice Chair provides opportunities to advocate for Codex all around the world.
My plan for the future is to continue to strengthen the participation of developing countries in Codex, to contribute to capacity building as well as follow the work of different codex committees hoping to reach the goal that everyone anywhere, anytime, has access to safe food.

Purwiyatno Hariyadi

I feel so honoured to be part of Codex family and elected as one of the three Codex Vice Chairs. My participation was very much jump-started by attending a workshop organized just before the 74th Session of the Executive Committee in 2017. Since then, I have had opportunities to attend and understand Codex work in detail by attending TFAMR5 (Jeju, Republic of Korea) and CCCF12 (Utrecht, The Netherlands). Being a Codex Vice Chair has also opened up more opportunities for me to promote the importance of Codex work and Codex Standards and I have participated in food safety events in Thailand and India this year. In the near future, I'm even more excited to contribute to the work of Codex, especially as CCCF13 (2019) is scheduled to be held in Indonesia. I am also looking forward to working together on the development and finalization of the Codex strategic plan 2020-2025.

Steve Wearne

It was a huge honour to have been elected as one of three Codex Vice Chairs and, under the leadership of Guilherme da Costa, we have collectively made a commitment to be closer and more accessible to Member countries. Although my contribution to Codex has been disrupted by illness in recent months, it was good to attend the meeting of the Executive Committee in 2017, as well as CCFH in Chicago. I look forward to making a full contribution to the work of Codex in the coming year, and foresee two particular priorities. First, the development and finalization of our strategic plan for 2020 and beyond, and the new focus we are proposing on advocating wider use of Codex standards by food businesses as well as Member countries. Second, maintaining momentum in our work on countering antimicrobial resistance, which remains one of the most significant risks for public health in the future.

Photos ©Bob Scott

Behind the WHO Five Keys to Safer Food

Why this campaign has worked for nearly two decades

by Françoise Fontannaz-Aujoulat and Kazuaki Miyagishima

Françoise Fontannaz-Aujoulat

Kazuaki Miyagishima

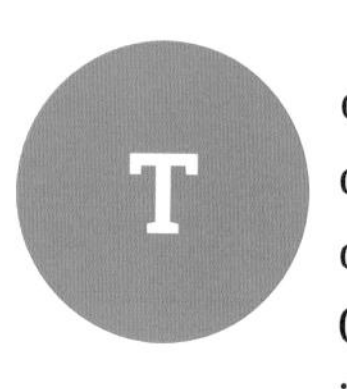

To show everyone what they can do to prevent foodborne diseases, the World Health Organization (WHO) introduced the 'Five Keys to Safer Food' in 2001.

Today as their 17th anniversary is celebrated, the Five Keys to Safer Food are recognized as one of the most successful campaigns and have been translated in 87 languages. Why did it work? Here are five secrets for its success.

Françoise Fontannaz, Technical Officer Risk Communication and Education Department of Food Safety and Zoonoses, World Health Organization

Kazuaki Miyagishima is Director of Food Safety and Zoonoses, World Health Organization

1. A simple message, easy to understand, adapt and adopt

The Five Keys to Safer Food are actionable because they are written in a plain language adapted to general public. As a result, their translation into a local language is very easy.

2. Evidence-based recommendations

Each Key invites the consumer to understand the 'why' behind the risks and what she or he can do about the risk. This increases the chances to trigger a real change in the consumer's behaviour.

3. A universal guidance relevant in all parts of the world

The Five Keys are relevant both in developing and developed countries, as common interventions to reduce foodborne pathogens are basically the same all over the world. The constant message of the Five Keys stands out in the flood of information of our age. The core recommendations of the Five Keys do not need be altered in different cultural settings.

4. Empowering consumers to get involved

Consumers want to do something to make their food safer, but few are aware of how. The Five Keys equip consumers with knowledge that drives them to request safe food and make informed choices. When they think they cannot act on risks, they feel powerless, start to ignore or exaggerate risks, and lose confidence in food systems.

Five Keys to Safer Food video available in 18 languages

5. Cross-sector integration in a wider context

When used in health promotion and education, the Five Keys can build dialogue and collaboration between different stakeholders from various sectors (health, agriculture, education, tourism, environment). All over the world, central and local authorities, schools, NGOs, food industry, consumers organizations use the Five Keys in health promotion campaigns, targeting health professionals, food handlers, food inspectors, adults and children, women and men.

Building on the success of the Five Keys to Safer Food, WHO produced the *Five Keys to Growing Safer Fruits and Vegetables* and the *Five Keys to Safer Aquaculture Products to Protect Public Health* to promote hygienic practices from farm to table. The Five Keys to Safer Food are now included in the recommendations for consumers to combat antimicrobial resistance.

The take-up by countries shows that simple messages coming from a trusted source can make a lasting difference.

On the 'future proof' track

Starting with a shared belief to keeping pace with technological innovation, Codex upholds its commitment to regulating the global food market

by Renata Clarke

A number of thoughts come to mind as preparations 'warm up' for this year's Codex Alimentarius Commission. The first of these relates to the very nature of the Commission: a body of people with diverse interests and perspectives coming together because of a shared belief in the need for common rules and common approaches for regulating the global food market. This 'shared belief' is essential but it is just the starting point of what it takes to make the Codex system work. There needs to be commitment to adequate preparation and a readiness to understand and to consider other countries' issues and interests. Ongoing work of the Codex Trust Fund and FAO's own capacity development work involves working with developing countries to strengthen the national processes that support solid Codex preparation. This is not only a message for developing countries: it is a perennially useful reminder to all to keep awareness high and to maintain commitment.

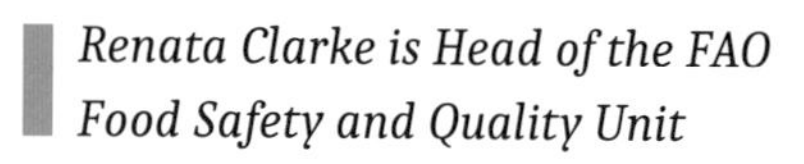

Renata Clarke is Head of the FAO Food Safety and Quality Unit

©FAO/Giuseppe Carotenuto

"We have to anticipate change and prepare for it. The stability and safety of our food supplies depend on it."

On a very positive note, there has been a lot of energy, enthusiasm and interest among Codex Members in participating in 'peer learning' under the Codex Trust Fund. Without a doubt, these interactions bring us all closer together and contribute to creating an environment that is conducive to constructive and respectful dialogue on food standards issues and for building global consensus.

My second thought is about change and innovation. No one can deny the rapid rate of change that is happening around us. I am told that the cameras on ongoing space missions that were launched about a decade ago are probably inferior to what most of us are walking around with in our telephones. Incredible! Such technological leaps are evident in many other areas as well that affect - directly and indirectly - food safety and our capacity to assess, manage and to communicate risk. Climate change, urbanization, changing political environments, economic transitions all drive change that is relevant to our capacity to effectively manage the safety of our food supplies.

Codex aspires to be 'future proof' and to be able to pick up and address change: meeting that aspiration depends on us. FAO has initiated work with countries to enable them to appropriately select and effectively apply 'foresight' techniques that allow them to gain intelligence on emerging food safety issues. There is a unique opportunity for this intelligence to be pooled and enhanced through Codex fora, such as the FAO/WHO Regional Coordinating Committee Meetings. The FAO sub-regional workshop on food safety foresight held in Nairobi in March 2018 generated much interest and is already prompting further action in some of the participating countries.

With the world changing so rapidly around us in so many ways, we cannot simply keep doing what we have always done and expect the same outcome: we have to anticipate change and prepare for it. The stability and safety of our food supplies depend on it.

Technology enhances our capacity to access, manage and communicate risk

An insider's story

A Codex Contact Point shares some tips, thoughts and her outlook for the future

Cassandra Pacheco is the Codex Contact Point for Chile

What does a routine day as a Codex Contact Point (CCP) look like?

Well, unexpected tasks come up that must be faced and resolved practically every day. But, in general, a CCP could comment on revisions by email, because that is the formal channel the Codex Secretariat uses to communicate with Member countries like Chile. So giving a timely response is an essential part of what a CCP does. In Chile, the role of the CCP today is complemented by the work we do as the regional coordinator for Latin America and the Caribbean (CCLAC).

What is the set-up of your team?

26 professionals work for the Chilean Agency for Food Quality and Safety (ACHIPIA). The Agency is divided into four major areas and the national and regional functions for Codex are housed in the International Affairs, whose focus is international cooperation and laws and multilateral organizations. We have a Coordinator, Diego Varela, and two other professionals, Constanza Vergara and Claudia Villarroel, besides me.

What is the next big challenge you see for Codex?

I believe that our biggest challenge now is elaborating a guideline for biopesticides, an issue that was addressed in the 50th session of the Codex Committee on Pesticide Residues. Chile together with the United States and India will preside over the electronic working group for the guideline. Chile raised this issue in the last Commission, where it received great support.

What challenges are you facing in your Codex work?

One of our challenging responsibilities is CCLAC coordination. On a daily basis we strive to make the region more active, cohesive and participatory in all Codex work. So, we are in constant communication with the region to work together, sharing our experience in Codex and supporting colleagues to carry out their work. This is how we have implemented cooperation projects that promote best practices and institutionalize Codex work in the region, thus demonstrating the importance of actively participating in international organizations.

What is special about being a regional coordinator?

I've never stopped to think about whether something makes a person special or not, for being in a specific position. What I can tell you is special about my work is the regular contact that one has with all parts of the world, understanding different realities, establishing friendships and working with so many diverse people and unique individuals is something that I appreciate. Having the opportunity to travel and tell Chile's experience in Codex is something that I love.

What advice would you give to the next regional coordinator?

An important aspect of this role is to be a very pro-active, to look for answers, since establishing objectives depends on how far you want to get with the work.

Photos ©ACHIPIA

When do you need INFOSAN?

For 15 years, this network has facilitated communication across borders, and between network members, during hundreds of food safety emergencies

by Peter Karim Ben Embarek and Carmen Savelli

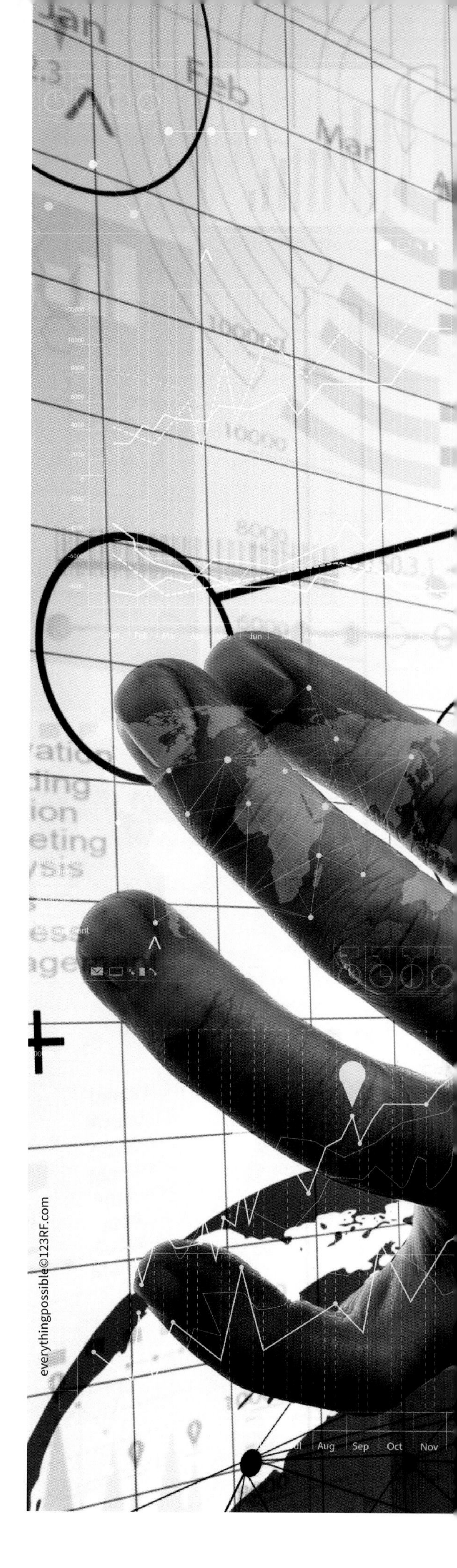

In our increasingly globalized world, it seems that more and more frequently we see largescale food safety incidents making headlines. These days, foodborne outbreaks easily span across multiple countries. A food product made in one country today might be consumed on the other side of the planet tomorrow. In the face of this reality, it is essential that international communication channels function rapidly when unsafe food enters the global market so that competent authorities may take rapid action to protect consumers from illness.

Peter K. Ben Embarek, Scientist, INFOSAN Management, World Health Organization

Carmen Savelli, Technical Officer, INFOSAN Secretariat, World Health Organization

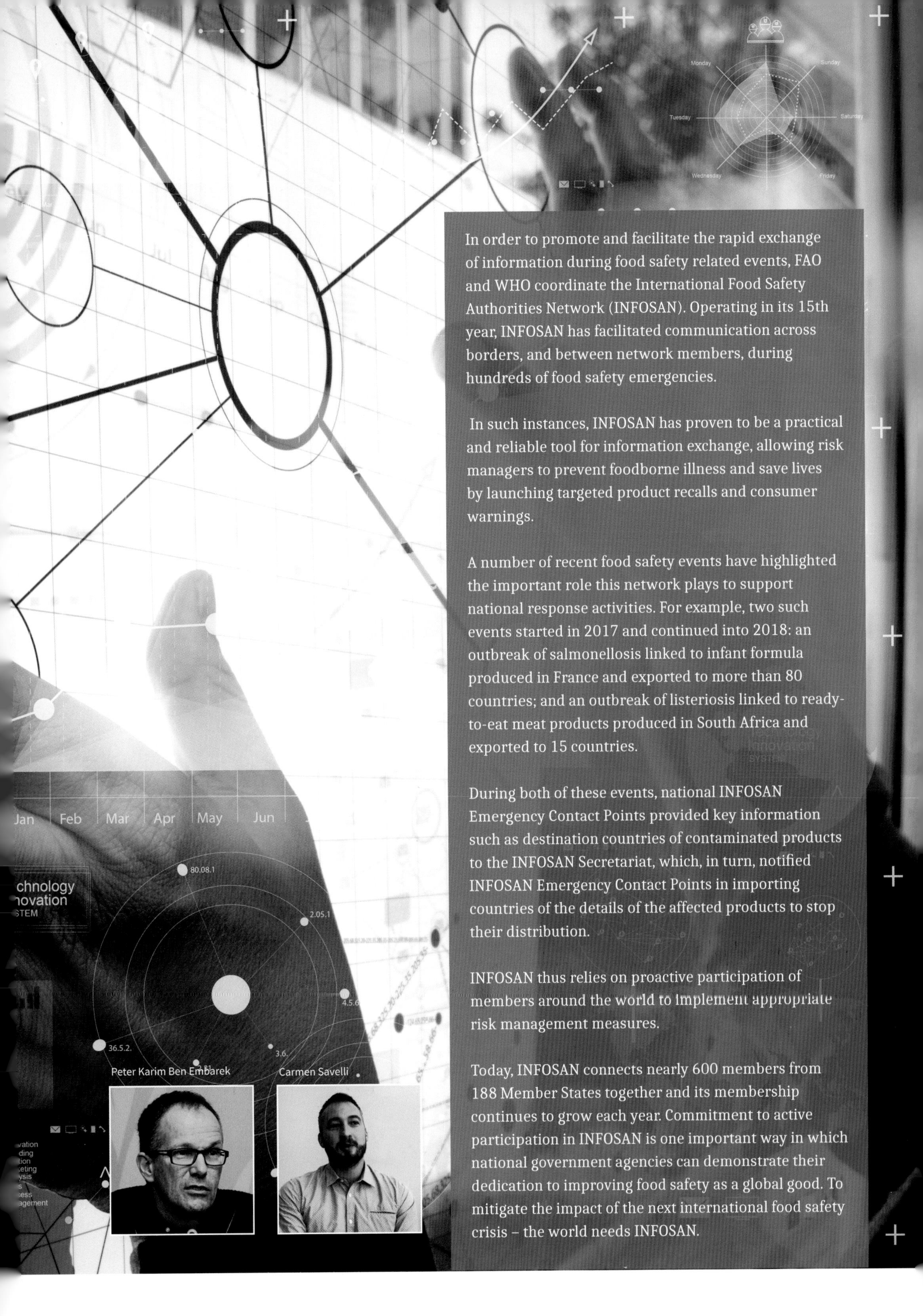

Peter Karim Ben Embarek

Carmen Savelli

In order to promote and facilitate the rapid exchange of information during food safety related events, FAO and WHO coordinate the International Food Safety Authorities Network (INFOSAN). Operating in its 15th year, INFOSAN has facilitated communication across borders, and between network members, during hundreds of food safety emergencies.

In such instances, INFOSAN has proven to be a practical and reliable tool for information exchange, allowing risk managers to prevent foodborne illness and save lives by launching targeted product recalls and consumer warnings.

A number of recent food safety events have highlighted the important role this network plays to support national response activities. For example, two such events started in 2017 and continued into 2018: an outbreak of salmonellosis linked to infant formula produced in France and exported to more than 80 countries; and an outbreak of listeriosis linked to ready-to-eat meat products produced in South Africa and exported to 15 countries.

During both of these events, national INFOSAN Emergency Contact Points provided key information such as destination countries of contaminated products to the INFOSAN Secretariat, which, in turn, notified INFOSAN Emergency Contact Points in importing countries of the details of the affected products to stop their distribution.

INFOSAN thus relies on proactive participation of members around the world to implement appropriate risk management measures.

Today, INFOSAN connects nearly 600 members from 188 Member States together and its membership continues to grow each year. Commitment to active participation in INFOSAN is one important way in which national government agencies can demonstrate their dedication to improving food safety as a global good. To mitigate the impact of the next international food safety crisis – the world needs INFOSAN.

Nuclear Codex

Collaborating with Codex on nuclear applications for food safety and control

by Carl Blackburn and Britt Maestroni

Working in partnership is a fundamental part of the '2030 Agenda', with Sustainable Development Goals (SDGs) that cover poverty, hunger, health and sanitation. Also, SDG 17 is devoted entirely to partnerships for the goals, a mark of joining hands to mobilize resources and strengthen implementation of sustainable development initiatives.

The Joint FAO/IAEA Division is itself a partnership, a programme shared between the FAO and the International Atomic Energy Agency (IAEA). At Codex, the Joint Division is the link to technical specialists in nuclear and related technologies as related to food safety and control. The Joint Division works with multiple partners to promote safe, secure and peaceful uses of nuclear technologies. Helping develop Codex food safety and quality standards related to nuclear technologies is a central part of our work. These comprise techniques to measure contaminants and residues in food; trace food origins; detect food fraud and adulteration; and utilize the energy from beams of ionizing radiation to maintain the quality of food and reduce the incidence of food poisoning. Expertise is also available to help countries prepare and respond to nuclear or radiological emergencies.

Codex standards and guidelines are THE basis on which the research and technical staff of the Joint Division and its associated FAO/IAEA Agriculture & Biotechnology Laboratories support Member countries through capacity building and training to ultimately enhance food safety, quality and control, to enhance trade and to protect consumers. Recent initiatives include promoting the consolidation of networks of analytical laboratories (for example in Latin America and the Caribbean); developing international guidance on radioactivity in food; coordinating research initiatives to develop integrated radiometric and complementary techniques for mixed contaminants and residues in foods; and developing field-deployable analytical methods to assess the authenticity, safety and quality of food.

Carl Blackburn and Britt Maestroni, Joint FAO/IAEA Division of Nuclear Techniques in Food and Agriculture

International Plant
Protection Convention

The International Plant Protection Convention (IPPC) has been contributing to food security, environmental protection and trade facilitation since 1952.

Food fit for the future

What it takes to keep our food safe

by Markus Lipp

As consumers, we may have a lot of different requirements or varying preferences for our ideal meals, but always, always, always, we expect our food to be safe - if it is not safe it is not food. Where pesticides are used, we expect their residues not to exceed the levels that are safe for humans. And when we eat peanuts we expect any potential amount of mycotoxins present to be low enough to not harm us or our children.

Markus Lipp, Senior Food Safety Officer, JECFA Secretariat

Rice fields on terraced of Mu Cang Chai YenBai

The Joint FAO/WHO Expert Committee on Food Additives (JECFA) is an international expert scientific committee that is administered jointly by FAO and WHO. It has been meeting since 1956, initially to evaluate the safety of food additives.

"Codex Alimentarius at its very heart provides guidance"

Those of us who have access to a local farm, may want to benefit from buying raw products directly from the producer. However, these raw food products still need careful preparation to render them safe. For example, while milk is very nutritious, raw milk will require pasteurization or boiling before consumption to prevent food poisoning.

Yet, not everybody may be able to enjoy having access to local farmers. By 2050, it is expected that more than half of the world population will be living in cities. To feed our cities, more food will need to travel longer distances, and be kept safe throughout the distribution process. This can require an uninterrupted cold chain from the farm to the refrigerator in our homes, as in the case of fresh milk or fresh fish, a logistical feat that is often taken for granted. In other cases, it can require more complex food processing, for example canning processes for vegetables and it likely requires more food to be packaged to maintain its safety and to preserve its quality.

Codex Alimentarius at its very heart provides guidance, and provisions that define the amount of contaminants and pesticide residues present that are safe for humans, enable the hygienic production of food and sets the appropriate levels for the safe use of preservatives and other food additives to ensure that we can enjoy their benefits safely – that we can enjoy our food safely.

Food is life and all life needs food to sustain itself. Food is a core part of our cultures and the subject to many passionate debates on how to best prepare a certain dish; tradition and family recipes are valued and guarded as a part of cultural identity. But underneath our cultural and dietary preferences and needs, underneath all the discussions, for our food to be safe, it must not contain harmful amounts of microbes or excessive levels of contaminants and residues and other substances – if it is not safe it is not food.

Part of the system

Effective preparation for successful participation in the international food trade

by Georgios Mermigkas, FAO Trade and Markets Division

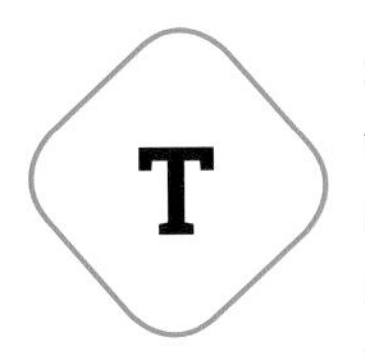

Trade can affect a wide number of economic and social variables, like the variety, quality and safety of food products, and the composition of diets. As such, it is directly linked to the efforts of governments to pursue food safety and food security objectives.

At the same time, governments use non-tariff measures (NTMs) to protect human, animal or plant health or to regulate technical characteristics of products. These NTMs can make it difficult for producers and exporters, especially in developing countries, to access other markets, if the requirements and standards in each market are different.

The system that comprises of the Codex Alimentarius Commission and the WTO SPS and TBT agreements allows governments to pursue their public policy objective of protecting public health while at the same time facilitating the international trade of safe nutritious food.

But for countries to reap the benefits of international trade, they have to be "part" of this system. They need to participate in the Codex processes and they need to be present and active in the SPS and TBT committee discussions.

Participation in the SPS and TBT committee work, is essential for countries to raise their concerns and discuss trade-related problems that arise from the implementation of SPS and TBT measures by their trading partners. The discussion allows for an honest exchange of opinions towards the resolution of potential conflicts while the presence of many experts from different countries, in the same place at the same time, gives the opportunity for informal resolutions of these concerns.

Another element is the creation of informal networks. Countries do not trade with themselves and maintaining open channels of communication between experts of different countries can facilitate trade and prevent concerns before them even being created.

However, successful participation in Codex and WTO related work requires effective preparation based on capacities and knowledge, which in its turn requires countries to invest in building these capacities as well as in involving all relevant stakeholders in a coordinated way.

The **Sanitary and Phytosanitary Measures (SPS) Agreement** sets out rules for food safety and requirements for animal and plant health.

The **Technical Barriers to Trade (TBT) Agreement** covers trade in all goods and aims to ensure that technical regulations, standards, and conformity assessment procedures are non-discriminatory and do not create unnecessary obstacles to trade.

Trade and food standards

When food standards and international trade work hand-in-hand, they help ensure food safety

Participation in the development of international food standards for trade is essential today more than ever. We met with WTO Director-General, Roberto Azevêdo during the launch of the *Trade and Food Standards* publication to talk about growth, development and drivers of change in the area of food regulation.

©FAO/Giulio Napolitano

Can you tell us about this joint WTO-FAO publication?

We did a fantastic job together with FAO and produced a publication precisely about the importance of standards for food safety and about facilitating food trade across borders. It is a very informative piece because it explains those two things, trade and food security, they do not clash, they actually reinforce each other. In the publication we show that if we have uniform international standards developed by Codex, those standards are going to make food safe for consumers and more affordable because countries themselves will not have to be individually burdened with the cost of developing those standards. The private sector, companies, will not have to develop those standards either, so the cost goes and everybody wins. Consumers have safer products at lower prices so this is the fantastic outcome we can expect from this publication.

Industrial port and container yard

Container Cargo freight ship

What are the compatibilities of the WTO and FAO frameworks that allow the two to come together to create an inclusive system for international food standards and trade?

FAO and WTO complement each other in a very nice way, because FAO working together with Codex has the expertise to develop the standards. They are the ones who can develop scientific standards which will be the benchmark for all the countries in the world. WTO has important agreements under its umbrella like the Sanitary and Phytosanitary Measures Agreement and the Technical Barriers to Trade Agreement which use these standards developed by Codex and FAO and consolidate them in obligations for Members to follow. A Member who complies with these standards is found to be automatically meeting its obligations with WTO. The two organizations therefore complement each other: one develops the standards and the second one has the discipline and the agreements which will implement the standards worldwide.

Roberto Azevêdo, World Trade Organization (WTO) Director-General

What challenges caused by trade interconnectedness do you see for food safety in the future?

Trade interconnectedness is a new element because as we connect more information across borders quickly and standards and concerns are more visible than ever before. Today if a product has a problem in one market, whatever that is for any particular reason, everyone else knows, and will know very quickly. So we have to ensure that we are in a position to respond to these things in a faster way. We have to make sure that standards are recognized, acknowledged and well-known across borders. We can also use technologies to develop safety and for testing and even to lower the cost of meeting those standards. I am pretty sure that we are going to have, more than ever before, safer food at more affordable prices with new technologies.

©FAO/Giulio Napolitano

Learn how FAO and WTO work together to facilitate trade based on internationally agreed food standards.
Publication now available in all UN official languages
Food and Agriculture Organization of the United Nations
WORLD TRADE ORGANIZATION
TRADE AND FOOD STANDARDS

Building Codex globally

In 2018 the following countries were successful in securing CTF2 funds for enhanced participation in Codex

Burkina Faso
Place Codex at the heart of national food safety issues. Enhance the participation of the National Codex Committee in the Codex Alimentarius.

Cabo Verde
Strengthen national Codex structures to enhance food safety and participation in Codex.

Guinea
Reinforce effective and sustainable participation in Codex.

Mali
Contribute to the improvement of the national infrastructure through increased participation of Mali experts in Codex activities.

Rwanda
Strengthen the National Codex Committee and increase the engagement of the country in Codex standardization activities.

Macedonia
Ensure well organized, knowledgeable, efficient and sustainable national Codex structures.

Bhutan-India-Nepal group project
Strengthen national Codex structures through efficient engagement of all stakeholders in Codex.

Honduras
Strengthen the structure of the National Committee and its consultation mechanisms, to improve the management of the work of the Codex Alimentarius.

20 years of risk analysis

How an idea ahead of its time is leading to a brighter future

by Steve Hathaway

The year, 1995. With the WTO SPS Agreement waiting in the wings in Geneva to be ratified, the CAC had commissioned a consultant report to introduce Members to a new way of thinking about setting standards using risk assessment. This new discipline was a core precept of the draft SPS Agreement and promised a science-based, transparent and fairer approach to setting international standards compared to the empirical and prescriptive approach that prevailed at the time. Most importantly, control measures were to be based on detailed knowledge on the likely risks to the consumer at the time of eating the food, not on the mere presence of the food safety hazard itself at different parts of the food chain prior to consumption.

It did not go well. With limited knowledge of the detail of the impending trade agreement and deep-seated suspicion of what would be a major change in Codex, the consultant was sent back to the drawing board. That consultant was myself.

The year, 1997. A reworked report focusing on risk management was presented to the CAC and this time, it was endorsed by acclamation. In the intervening years, the WTO SPS Agreement had been signed and Members had begun to explore the generic risk analysis provisions therein and had recognised their critical value in improving the utility and uptake of Codex standards. This endorsement began a wave of irrevocable change in Codex, beginning with a decade of work to codify the principles of risk assessment (the science), risk management (the decision making on the control measures) and risk communication (engaging with all stakeholders) in the Procedural Manual. This body of work remains robust today. It was followed by a further decade of work by relevant Codex committees to incorporate risk analysis principles and specific methodologies in a new wave of risk-based standards and this more specific work continues apace today.

What is the difference between hazard and risk?

A hazard is something that can be dangerous, or harmful to your health. A big wave can potentially harm you but if you don't go in the water there is no risk to your health.

Simple is not stupid and complex is not always correct

Steve Hathaway is the Director of Science and Risk Assessment, Ministry of Primary Industries of New Zealand

In the 1990s, consumers (and therefore governments) were much more worried about chemicals in the food supply than microorganisms. This was a paradoxical situation for regulators, given that deterministic if conservative methodologies were available for setting food safety limits; these were already somewhat risk-based and there was no epidemiological evidence of public health problems from chronic exposures. On the other hand, food-borne illness due to micro-organisms was evident on a huge scale in many countries. Despite this, food safety standards were mostly based on the presence of hazards in foods rather than on risks to consumers, and were often inflexible and unnecessarily burdensome on industry.

This scenario was ripe for exploration. Risk assessors largely accepted the deterministic methodologies in place for chemicals but seized on the opportunity to develop risk-based standards for microbial pathogens. However, they largely underestimated the data gaps that would need to be filled to generate a reasonably certain risk estimate for a specific hazard and a specific exposure pathway. Thus the early years were somewhat scarred by massive, multi-year efforts that provided highly complex and elegant probabilistic models but these were often too uncertain in their risk estimates to be of value to decision-makers, or they were so tied to a specific situation that they were not useful for more generic standard-setting.

The 2000s brought a raft of experience in applying risk analysis at the national level and the advent of a toolbox of approaches to turn available data into risk-based standards. “Simple is not stupid and complex is not always correct” was a phrase coined by Marcel Zwietering that described a more practical way of thinking about assessing risk and this persists today. This experience has influenced international standard setting to the extent that national governments now often rely on Codex for their national risk-based standards. Further, consumers (and governments) are now more rightly focused on microbial hazards rather than chemical hazards as the greatest source of food-borne illness.

The future is bright for continued and more effective use of risk analysis in setting of Codex standards for food in international trade. The body of risk analysis principles that was painstakingly put together in earlier years has stood the test of time and has also found its way into the national legislation of many countries.

Despite this guidance, the diversity of food chains and national diets around the world means that there will always be challenges ahead in setting quantitative risk-based standards for food in trade that suit all. However, those smart people enshrining risk analysis principles in the WTO SPS Agreement had already thought of this and thereby included a principle that allows national governments to put in place more stringent measures than those of Codex where they are properly justified by risk assessment and consumer protection. A bright future indeed!

Food and Agriculture Organization of the United Nations
UNDERSTANDING CODEX
CODEX ALIMENTARIUS
World Health Organization
ow also available
Arabic, Chinese
ench, Russian
d Spanish

Observing Codex

We met with Martin Slayne to get an industry perspective from within Codex

How do you see Observer organizations contributing to the work of Codex?

It is important to focus on the scope for harmonization, taking into account valuable information from all stakeholders, including the constraints of commercial realities and best practices, helping to reduce costs and enable trade. Non-governmental organizations equally want to ensure that food is safe, and that we are protecting public health, based on all relevant information, to shape regulations around the exposure risk, best science, best practices and reality for what is globally achievable. It is important to communicate and help more people understand the value of Codex, encouraging all stakeholder groups to come to Codex as a first place to help inform standard-setting.

©Bob Scott

Technologies, block chain, science are shaping food systems. Where do Codex stakeholders fit in this scenario?

Stakeholders generally see the importance and value of harmonizing standards. There is a need to communicate this value more broadly. Some non-governmental stakeholders are skeptical, thinking they cannot influence Codex standards. There is an opportunity to correct these perceptions and encourage more sharing of information into the decision-making. The global harmonization of standards helps trade, helps producers and helps producing-countries.

©Bob Scott

Paulo Gonçalves@123RF.com

In your view, how can we more effectively engage different stakeholders?

As a starting point, it is valuable to encourage more voices and sharing from Observer delegations, maybe not framing them as 'Observers' but as 'Non-governmental delegations', more clearly as participants than just observers. For example, Member countries always come first in the meetings and sometimes the non-governmental delegations are either not forthcoming with comments or their comments are not considered further. Encouraging non-governmental delegations to share more information, particularly on practical matters, is an opportunity for Codex. Sharing information is about finding a better solution for everyone.

Let's talk about the process to determine 'As Low As Reasonably Achievable' for chemical contaminants, commonly referred to as ALARA. Can ALARA determination be quicker, to anticipate individual Member countries already taking action? If Codex can demonstrate that it can effectively align Members on the principles of applying global ALARA, to different scenarios of standard-setting, and if Member countries are motivated to align on global ALARA, then industries and producers will be able to further help governments come to harmonized solutions leveraging wider data.

The world is now changing faster than ever before. Should Codex change as well?

Across the globe initiatives on food labelling are highly active, to provide consumers with information about the food they eat, yet the Codex Committee on Food Labelling (CCFL) meets only every 18 months and cannot keep up with the food labelling landscape. This committee could be more dynamic, addressing hot topics, such as front-of-pack labels and allergen warning labels. Maybe include more global review education at the Committees - for example, on allergen warning labels invite support group perspectives and education on why allergens are a global concern and help get to commonality on labelling.

Overall, there is an opportunity for Codex to actively promote its valuable forum for non-governmental organizations to more actively contribute, not framing them just as 'Observers', and help us all work together towards securing simpler, widely-executed, global harmonized standards.

Martin Slayne is Global Head, Scientific and Regulatory Affairs at The Hershey Company, a member of the International Council of Grocery Manufacturers Associations (ICGMA), which contributes to the work of the Codex Alimentarius Commission and nine Codex committees.

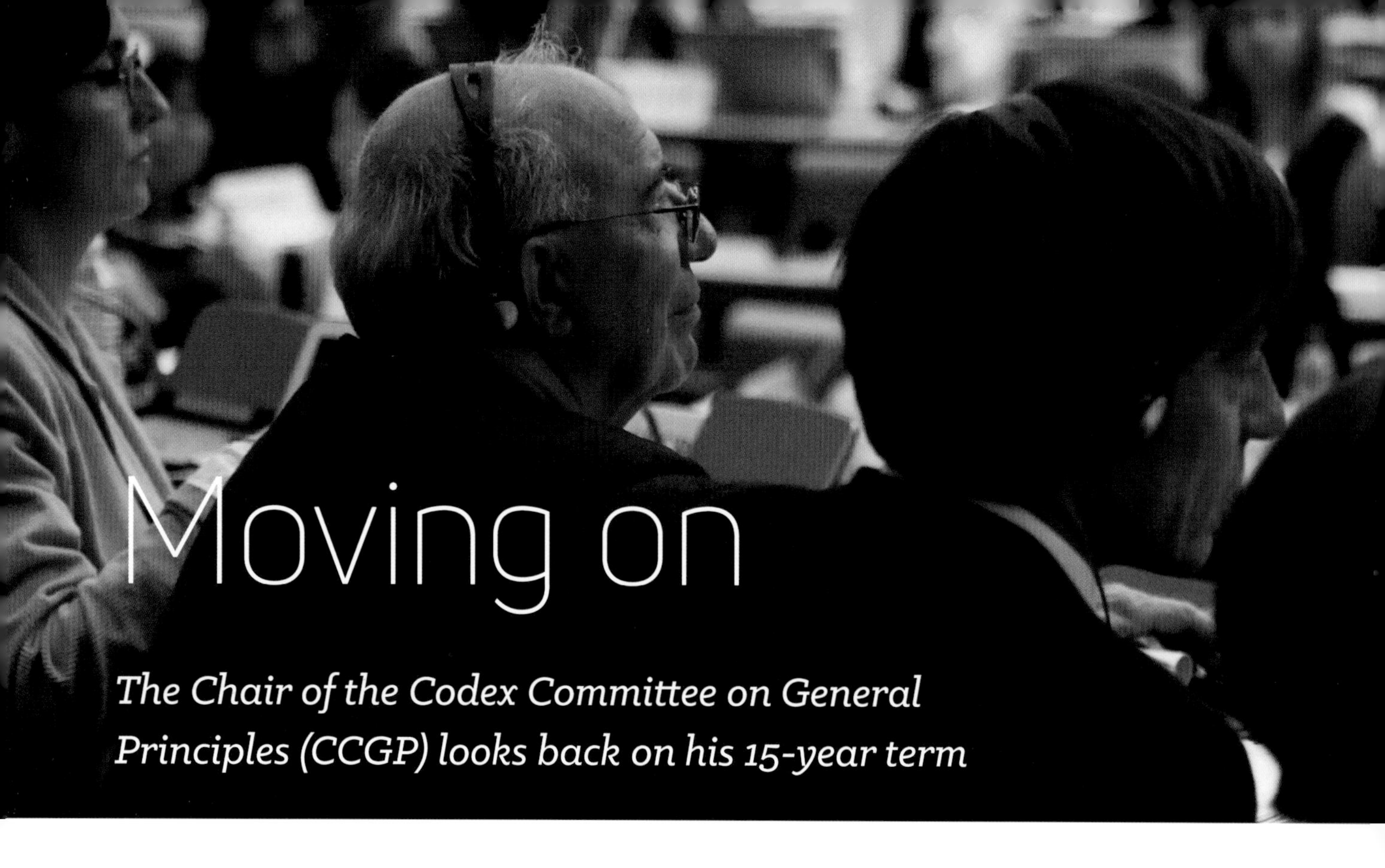

Moving on

The Chair of the Codex Committee on General Principles (CCGP) looks back on his 15-year term

Michel Thibier chaired the Codex Committee on General Principles (CCGP) from 2003 to 2018. He retired this year and spoke to us about Codex.

When did you start in Codex?

As the French Ministry of Agriculture Permanent Representative to FAO (2000-2002), my first interaction with Codex staff was in Rome and, as my background was that of a veterinarian, I followed quite carefully, what was going on in this area.

Back in France, as Director General of Education and Research of the Ministry of Agriculture in Paris, I was asked, knowing Codex well, to chair CCGP in 2003. I held this position until 2018, having done my time so to speak...

What would you think is critical to make the meetings more efficient?

Surprisingly maybe, the first critical points to me are the good relationship and close interaction both within the team of the focal point here in Paris and those with the Secretariat in Rome. I was quite aware of this importance to start with in Rome and confirmed through further experience. I would like to take this opportunity to thank very much all the members of the French team - Roseline, Sophie and Louise together with all the other members. I would also like to mention that I appreciated the good relationship we have always had with the Secretariat in Rome, Tom and his predecessors as well as all his team. Good relationships are definitely critical to achieving a lot at the meetings.

What have been the highlights of your (Codex) career?

What comes to mind are the many cases in which there was no consensus at first but then, after some time and many discussions in the room or out of the room in an appropriate way, we eventually succeeded in finding compromises acceptable to everyone. Knowing many delegates over the years, we could trust each other and move forward in a positive direction. I cannot stress enough the importance of having committee chairs for a period long enough to establish good relationships with the delegates, who generally participate over several consecutive years.

What is the role of CCGP in the current Codex system?

I have seen that the role of the CCGP, a special type of committee indeed, brings a most essential support to the governance of Codex, CAC and CCEXEC.

To me, the CCGP is a transparent, inclusive, open forum, which aims, in particular, to deal with amendments to the procedural manual. It should constitute more

©Bob Scott

broadly a committee of reflection on all the strategic subjects of Codex (review of the conduct of work of Codex, link with other international organizations, strategic plan.) So, such a committee is a good place to discuss in an inclusive manner matters related to the efficiency of the work to be done by Codex.

What are the big issues Codex will have to tackle in the next 10 years?

To me there are many issues of different levels to tackle in the 10 years ahead. Operationally, one is that the work done is too slow, much too slow and it affects the credibility and the efficiency of the work itself. New technologies should be used much more widely, mitigating the costs of running Codex while speeding up the processes.

In the short term, Codex has to declare a clear position on how to deal with current sensitive issues worldwide. Issues such as the SDGs, antimicrobial resistance, the 'legitimate factors' issue in the Procedural Manual in the *Principles Concerning the Role of Science in the Codex Decision-Making Process* and other topics, for example, emerging risks to food safety.

In the medium term: Codex relies on the Governance of WHO and FAO, and this may have some advantages but certainly some disadvantages. For example, I am a bit puzzled by the lack of sufficient knowledge of all the innovations and outcomes of research, which seem not to reach the Members sitting in the committees. How could this be solved?

In terms of vision, it does not seem to me that the Governance of Codex has thought deeply about what Codex should be in 10 years' time. Will it still be necessary? Who will be the beneficiaries?

Michel Thibier, Chair of the Codex Committee on General Principles (2003-2018)

©Bob Scott

When #Codex trends on Twitter

Codex committees engage in virtual conversation limited to 280 characters, social media style

Codex Committee on Food Import and Export Inspection and Certification Systems (CCFICS)
Australia
Come to Brisbane, Australia in October 2018 to help shape CCFICS guidance on the use of systems equivalence.

Codex Committee on Residues of Veterinary Drugs in Foods (CCRVDF)
United States of America
Veterinary drugs are critical tools in the production of a safe and abundant food supply. Safety of the veterinary residues in food is assured by Codex standards.

Codex Committee on Pesticide Residues (CCPR)
China
Relying on your work, offering you services, CCPR sets MRLs and designs guidance documents that facilitate fair trade in food and ensure the safe use of pesticides to protect consumer health as part of the Codex Alimentarius international food standards.

Codex Codex Committee on Nutrition and Foods for Special Dietary Uses (CCNFSDU)
Germany
Did you know that CCNFSDU contributes to a healthy start in life? We set the scene for nutritious foods for infants and young children and take care of young and older patients with special dietary needs as well.

FAO/WHO Coordinating Committee for Latin America and the Caribbean (CCLAC)
Chile
Codex is about creating opportunities, opportunities for everyone to access safe food, and opportunities for large, medium and small stakeholders, to take part in the global food trade.

FAO/WHO Coordinating Committee for Europe (CCEURO)
Kazakhstan
CAC40 proposed to establish a World Food Safety Day that would take place ubiquitously and annually on 7 June. Even though the date isn't officially approved yet, we appeal to people - and not only this day - to make their contribution to a safer and healthier world.

Ad hoc Codex Intergovernmental Task Force on Antimicrobial Resistance (TFAMR)

Republic of Korea

Our friends & neighbours are dying from antimicrobial resistant infections. As micro-organisms do not respect borders, all of us should work together within the 'One Health' paradigm to combat AMR. Join our efforts to mitigate AMR in the food chain, search 'Codex TFAMR' on the web.

Codex Committee on Spices and Culinary Herbs (CCSCH)

India

The Codex Committee on Spices and Culinary Herbs has proven to be really efficient finalizing 3 Codex standards in just 3 sessions.

Codex Committee on General Principles (CCGP)

France

Since 1965 the Codex Committee on General Principles has ensured that Codex procedures and working practices contribute to the international food trade through modern, appropriate and legitimate standards.

FAO/WHO Coordinating Committee for Africa (CCAFRICA)

Kenya

A common African proverb "If you give bad food to your stomach, it drums for you to dance" is associated with food safety and can be used in the region to create awareness and rally countries to adopt food safety standards set by the Codex Alimentarius Commission.

Codex Committee on Fresh Fruits and Vegetables (CCFFV)

Mexico

Did you know that CMCAC is the Mexican Codex Alimentarius Committee? It´s supported by different government offices, chambers and associations & its primary objective is to establish the positions of Mexico for international standardization to safeguard consumer protection and fair trade.

Codex Committee on Contaminants (CCCF)

The Netherlands

Important in CCCF's work is the development of good production practices to prevent or reduce contamination of food and feed with chemical contaminants. These Codes of Practices (COPs) help producers in the production of safe food.Select 'CCCF' in the list of COPs to find out.

@FAOWHOCodex

Canada

Did you know that Canada co-hosted the 44th Session of the Codex Committee on Food Labelling with Paraguay in October 2017 where the Committee identified for discussion new and emerging food labelling issues?

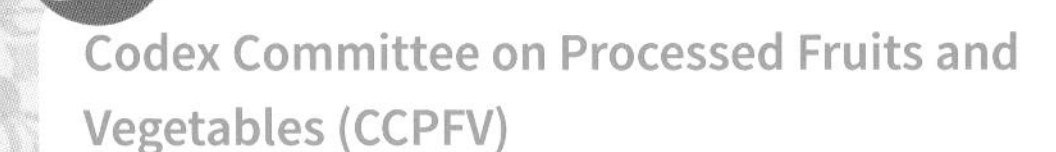

United States of America

Taking advantage of Codex's electronic forum and user group capability has allowed CCPFV to continue its current work cost-effectively. Thank you to the Members participating and to the Codex Secretariat for making the resources available!

Codex Committee on Methods of Analysis and Sampling (CCMAS)

Hungary

Did you know that the Codex Alimentarius has a collection of methods of analysis and sampling? By consulting CXS 234-1999, you can find all the methods endorsed by the Codex Committee on Methods of Analysis and Sampling (CCMAS) in one single place.

FAO/WHO Coordinating Committee for Asia (CCASIA)

India

Asia is known to be a place of rich culture and heritage, including a trove of delectable street foods. We encourage use of the Regional Code of Hygienic Practice for Street-Vended Foods so that street food will not only be tasty but safer and healthier too.

Vanuatu

We look forward to a successful workshop in Vanuatu in 2018 and an important regional meeting in 2019.

FAO/WHO Coordinating Committee for Near East (CCNE)

Iran

If Members of the near east region have effective communication and cooperation with each other regarding critical and emerging issues many problems will be solved and Codex goals will be achieved.

Codex Committee on Food Hygiene (CCFH)

United States of America

If you like to work with dozens of other countries that communicate in at least three languages to achieve consensus in food safety, CCFH is for you. We draft documents based on solid science that support the fair trade of safe food. And we have fun doing our job!

Codex Committee on Food Additives (CCFA)

China

On a foundation of scientific advice and consensus and the 'one CCFA' principle proposed by Professor Chen, we will be able to set standards to enhance food safety, to protect consumers worldwide and to remove barriers to international food trade.

Codex Committee on Fats and Oils (CCFO)

Malaysia

The work of CCFO will remain relevant and challenging, due to increased demand for healthier fats and oils. I am confident with the prevailing spirit of cooperation and goodwill among delegates, the work of CCFO will continue to progress in a transparent and efficient manner.

The burden of foodborne diseases is substantial

Every year foodborne diseases cause:

almost **1 in 10** people to fall ill

33 million healthy life years lost

Foodborne diseases can be deadly, especially in children <5

420000 deaths

Children account for **1/3** **of deaths from foodborne diseases**

Foodborne diseases are caused by types of:

Bacteria

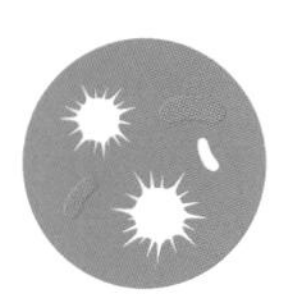

Viruses

Parasites

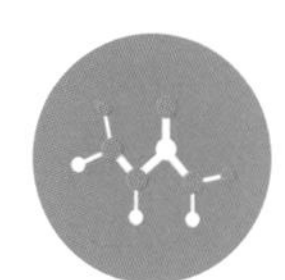

Toxins

Chemicals

Some of these are a public health concern across all regions
Others are much more common in middle- and low-income countries

FOODBORNE DISEASES ARE PREVENTABLE.
EVERYONE HAS A ROLE TO PLAY.

or more information: **www.who.int/foodsafety**

SafeFood

urce: WHO Estimates of the Global Burden of Foodborne Diseases. 2015.

Codex is people

A portrait of Lingping Zhang, Food Standards Officer

Lingping Zhang is recognizable in Codex with her friendly smile and can-do attitude. As a member of the truly global Codex Secretariat since 2014, Lingping hales from the People's Republic of China, where she previously worked on the development and implementation of a comprehensive national food-standards programme and food-control system to protect consumers throughout the Hong Kong Special Administrative Region – a bustling global crossroads that imports 90% of its food. A graduate of China's Peking Medical University, Lingping long aspired to devote herself to protecting public health more broadly.

What does your normal day look like?

"A normal Codex day?" – her easy laugh appears – "No such thing exists!" Indeed, working as a staff member of Codex, mother of a teenage boy and with a range of other responsibilities, Lingping has plenty to keep her busy – and in very different ways. Maintaining a responsive presence with Codex stakeholders across the world, coordinating working documents and practical arrangements for committees and working groups, representing Codex at UN and partner meetings, advising colleagues and contributing to crosscutting initiatives – all of this can be thirsty work, making Lingping recognizable wherever she may be in the world with her trusty flask of herbal tea in hand. In her spare time, she is busy cooking and dealing with the many unexpected challenges and adventures involved in raising a teenage son!

©Bob Scott

How do you manage your responsibilities and keep on smiling – is there a secret ingredient in that herbal tea?

"Staying hydrated is essential to keep your brain running efficiently and meet challenging demands!" Such is the sort of common-sense perspective Lingping shares with her colleagues, with Codex teams often working late into the night. "Thorough preparation, open communication and keeping things in perspective, this is my recipe for being a professional – oh, and eating well: good food is essential, and better still if it's spicy – !"

So there we have it, what makes an effective Food Standards Officer: technical expertise and attention to detail, with a healthy measure of tireless dedication and a pinch of humour. By demonstrating these attributes, Lingping exemplifies effective, results-oriented technical cooperation that responds to all stakeholders' needs. "Bearing in mind the concerns and motivations we all share as individuals and as households is no reason to limit our effectiveness: on the contrary, enjoying the human dimension of our exchanges as we negotiate technical agreements to protect health and facilitate trade, balancing the interests of all stakeholders, this is how Codex can continue to produce unparalleled global public goods for the 50 years to come". Her partners throughout the Codex community couldn't agree more and are glad to have her on board.

"Thorough preparation, open communication and keeping things in perspective, this is my recipe for being a professional"

Improving cooperation with standards

The Codex Secretariat recently looked at how to enhance its cooperation with Observers that engage in the development of standards or methods

There is a market for non-standardized products – just think of the number of different plugs you need to attend Codex meetings. While the absence of international standards may lead to economic opportunity in certain areas or individual cases, history has taught numerous lessons of how standards helped facilitate everyday life, increase safety and rationalize operations. Standards are an instrument of self-management and serve the interest of the general public. In the area of food, standards and regulations can be traced back to the times of the earliest societies (the ancient Romans for instance had standard practices for preserving their food under olive oil or salt). it was only after World War II that activity in international standardization increased significantly and international organizations with standard-setting organizations such as the International Organization for Standardization (ISO, 1947), the United Nations Economic Commission for Europe (UNECE, 1947) and the Codex Alimentarius Commission (1962) were founded due to the increased need for cooperation, in particular in the rapidly expanding international food trade market.

Today standards build an invisible infrastructure and can be found in every aspect of our daily life. They vary greatly; they may take the form of Codes of practice, symbols, methods or guides), fulfil different needs and cover different aspects of a certain product or service (e.g. its safety, quality, testing, comparability, economic/social/environmental management). The Codex Alimentarius Commission, developing standards in the area of food safety and quality, considers coordination with other international organizations an important part of its mandate.

©FAO/Alessia Pierdomenico

Today standards build an invisible infrastructure and can be found in every aspect of our daily life

©FAO/Shah Marai

Some of the organizations Codex talked to, such as the Organisation for Economic Co-operation and Development (OECD) and International Organization for Standardization (ISO), are well known due to their wide field of activities. Others may be lesser known because they focus on one specific sector or issue or are smaller in size, such as the Organisation Internationale de la Vigne et du Vin (OIV) and the International Federation of Organic Agriculture Movements (IFOAM).

However, as far as the development of standards is concerned, all these organizations rely on the consensus of their members to agree on what is best for intended users of a certain product, service or process. Another common feature is that all organizations working on international standards seek to contribute to trade facilitation, safety enhancement, the protection of health or a combination of those objectives, which is also why many international standard setters are currently assessing their contribution to the UN Sustainable Development Goals (SDGs).

©FAO/Ezequiel Becerra

©Flooded Cellar/Sue Price

The review conducted by the Codex Secretariat found that international standard-setting organizations face similar challenges to Codex in measuring, monitoring and reporting on the uptake of their standards and there are several opportunities to improve engagement and dialogue with benefits for both parties. Codex will eagerly continue to work on improving its cooperation and coordination mechanisms with other international standard setters in the years to come. Ideas that came up when talking to the organizations included how different organizations cross reference each other; collaboration on SDGs and major issues such as food fraud, or perhaps developing joint capacity-building opportunities.

This year in CODEX

JULY 2017

CAC 40 CODEX

Codex Alimentarius Commission

Geneva, Switzerland

START

SPETEMBER 2017

CCEXEC 74

Executive Committee of the Codex Alimentarius Commission

Rome, Italy

OCTOBER 2017

CCFFV 20

Codex Committee on Fresh Fruits and Vegetables

Kampala, Uganda

CCFL 44

Codex Committee on Food Labelling

Asunción, Paraguay

NOVEMBER 2017

CCFH 49

Codex Committee on Food Hygiene

Chicago, United States of America

TFAMR 5

Ad hoc Codex Intergovernmental Task Force on Antimicrobial Resistance

Jeju, Republic of Korea

DECEMBER 2017

CCNFSDU 39

Codex Committee on Nutrition and Foods for Special Dietary Uses

Berlin, Germany

FEBRUARY 2018

WORKSHOP
Codex Tools
Nairobi, Kenya

WORKSHOP
Host Secretariats
Paris, France

WORKSHOP
Codex Tools
Dakar, Sénégal

CCPFV
Codex Committee on Processed Fruits and Vegetables
by correspondence

Codex Committee on Food Additives
Xiamen, Fujian Province, China

CCCF 12

Codex Committee on Contaminants in Foods
Utrecht, Netherlands

WORKSHOP
Policy Convergence and Standard Implementation
Quito, Ecuador

MARCH 2018

APRIL 2018

CCPR 50

Codex Committee on Pesticide Residues
Haikou, China

CCRVDF 24

Codex Committee on Residues of Veterinary Drugs in Foods
Chicago, United States of America

CCMAS 39

Codex Committee on Methods of Analysis and Sampling
Budapest, Hungary

WORKSHOP
Codex Tools
Asunción, Paraguay

MAY 2018

JUNE 2018

CCEXEC 75

Executive Committee of the Codex Alimentarius Commission
Rome, Italy

CAC 41
CODEX

Codex Alimentarius Commission
Rome, Italy

JULY 2018

List of Members

- Afghanistan
- Albania
- Algeria
- Angola
- Antigua and Barbuda
- Argentina
- Armenia
- Australia
- Austria
- Azerbaijan
- Bahamas
- Bahrain
- Bangladesh
- Barbados
- Belarus
- Belgium
- Belize
- Benin
- Bhutan
- Bolivia (Plurinational State of)
- Bosnia and Herzegovina
- Botswana
- Brazil
- Brunei Darussalam
- Bulgaria
- Burkina Faso
- Burundi
- Cabo Verde
- Cambodia
- Cameroon
- Canada
- Central African Republic
- Chad
- Chile
- China
- Colombia
- Comoros
- Congo
- Cook Islands
- Costa Rica
- Croatia
- Cuba
- Cyprus
- Czech Republic
- Côte d'Ivoire
- Democratic People's Republic of Korea
- Democratic Republic of Congo
- Denmark
- Djibouti
- Dominica
- Dominican Republic
- Ecuador
- Egypt
- El Salvador
- Equatorial Guinea
- Eritrea
- Estonia
- Eswatini (Kingdom of)
- Ethiopia
- European Union
- Fiji
- Finland
- France
- Gabon
- Gambia
- Georgia
- Germany
- Ghana
- Greece
- Grenada
- Guatemala
- Guinea
- Guinea-Bissau
- Guyana
- Haiti
- Honduras
- Hungary
- Iceland
- India
- Indonesia
- Iran (Islamic Republic of)
- Iraq
- Ireland
- Israel
- Italy
- Jamaica
- Japan
- Jordan
- Kazakhstan
- Kenya
- Kiribati
- Kuwait
- Kyrgyzstan
- Lao People's Democratic Republic

On February 20th 2018, the Government of Timor-Leste became the 188th Member of the Codex Alimentarius Commission.

- Latvia
- Lebanon
- Lesotho
- Liberia
- Libya
- Lithuania
- Luxembourg
- Madagascar
- Malawi
- Malaysia
- Maldives
- Mali
- Malta
- Mauritania
- Mauritius
- Mexico
- Micronesia (Federated States of)
- Mongolia
- Montenegro
- Morocco
- Mozambique
- Myanmar
- Namibia
- Nauru
- Nepal
- Netherlands
- New Zealand
- Nicaragua
- Niger
- Nigeria
- Norway
- Oman
- Pakistan
- Panama
- Papua New Guinea
- Paraguay
- Peru
- Philippines
- Poland
- Portugal
- Qatar
- Republic of Korea
- Republic of Moldova
- Romania
- Russian Federation
- Rwanda
- Saint Kitts and Nevis
- Saint Lucia
- Saint Vincent and the Grenadines
- Samoa
- San Marino
- Sao Tome and Principe
- Saudi Arabia
- Senegal
- Serbia
- Seychelles
- Sierra Leone
- Singapore
- Slovakia
- Slovenia
- Solomon Islands
- Somalia
- South Africa
- South Sudan
- Spain
- Sri Lanka
- Sudan
- Suriname
- Switzerland
- Syrian Arab Republic
- Tajikistan
- Thailand
- The former Yugoslav Republic of Macedonia
- Timor-Leste
- Togo
- Tonga
- Trinidad and Tobago
- Tunisia
- Turkey
- Turkmenistan
- Uganda
- Ukraine
- United Arab Emirates
- United Kingdom
- United Republic of Tanzania
- United States of America
- Uruguay
- Uzbekistan
- Vanuatu
- Venezuela (Bolivarian Republic of)
- Viet Nam
- Yemen
- Zambia
- Zimbabwe

How the Codex Alimentarius is funded

Public international standards depend on the support of governments, trusts and individuals

hile the Codex programme budget of around USD 8.8 million for 2016-2017 makes up the largest share, the Codex system additionally relies on extra-budgetary support of Members, contributions of trusts and individuals, namely: in-kind contributions of host countries to Codex Committee and Task Force sessions and the working group meetings established by them; the provision of scientific advice from experts of FAO and WHO; The FAO/WHO Codex Trust Fund 2; FAO and WHO capacity building projects and events related to Codex at national and regional levels as well as extra-budgetary funding of staff in the Codex Secretariat by individual Codex Members.

Funding in the biennium 2016-17 helped finalize 17 new standards and over 50 standard revisions

The Codex Secretariat spent USD 8.738 million of the total budget of USD 8.789 million for 2016-17. The largest share went to covering the costs of 12 fixed term staff members (47%), followed by translation, interpretation and printing services for Codex meetings (18%) and consultancy as well as contracts with external service providers (16%).

Meetings of the Codex Alimentarius Commission and its Executive Committee amounted to 20% of the budget. Other important activities in the 2016-17 biennium included the development of guidance documents; workshops and information activities for different Codex stakeholders; costs related to the Codex website and several electronic tools used throughout the standard development process.

List of standards proposed for adoption at CAC41

Codex Committee on Fresh Fruits and Vegetables (CCFFV20)

- Standard for aubergines

Codex Committee on Food Labelling (CCFL44)

- Draft Revision of the General Standard for the labelling of prepackaged foods: date marking

Codex Committee on Food Hygiene (CCFH49)

- Editorial amendments to the Code of hygienic practice for low-moisture foods

Codex Committee on Contaminants in Foods (CCCF12)

- MLs for lead in selected commodities (revision of MLs/revocation of corresponding MLs/amendments to MLs)
- MLs for cadmium in chocolate containing or declaring ≥ 50% to < 70% total cocoa solids on a dry matter basis; and chocolate containing or declaring ≥ 70% total cocoa solids on a dry matter basis
- MLs for tuna, alfonsino, marlin and shark
- Code of practice for the prevention and reduction of dioxins, dioxin-like PCBs and non dioxin-like PCB contamination in food and feed

Codex Committee on Food Additives (CCFA50)

- Specifications for the identity and purity of food additives
- Revision of the Class Names and the International Numbering System for Food Additives

- Revised food-additive provisions of the GSFA in relation to the alignment of the annexes on canned mangoes, canned pears and canned pineapples of the Standard for Certain Canned Fruits
- Standard for Certain Canned Fruits
- Standard for Canned Salmon
- Standard for Canned Shrimps or Prawns
- Standard for Canned Tuna and Bonito
- Standard for Canned Crab Meat
- Standard for Canned Sardines and Sardine-Type Products
- Standard for Canned Finfish
- Standard for Salted Fish and Dried Salted Fish of the Gadidae Family of Fishes
- Standard for Dried Shark Fins
- Standard for Crackers from Marine and Freshwater Fish, Crustacean and Molluscan Shellfish
- Standard for Boiled Dried Salted Anchovies
- Standard for Salted Atlantic Herring and Salted Sprat
- Standard for Sturgeon Caviar
- Standard for Fish Sauce
- Standard for Smoked Fish, Smoke-Flavoured Fish and Smoke-Dried Fish
- Revised food-additive sections of Standards for Milk Powders and Cream Powder
- Revised food-additive sections of Standard for Blend of Skimmed Milk and Vegetable Fat in Powdered Form
- Revised food-additive sections of Standard for Edible Casein Products
- Revocation of relevant food-additive provisions from the Standards for Mozzarella (CXS 262-2006), Cottage Cheese (CXS 273- 1968), Cream Cheese (CXS 275-1973), Fermented Milks (CXS 243-2003) and Dairy Fat Spreads (CXS 253-2006).
- Revocation of the relevant food-additive provisions for sodium sorbate (INS 201) from the Standards for Instant Noodles (CXS 249-2006), Fermented Milks (CXS 243-2003), Dairy Fat Spreads (CXS 253-2006), Mozzarella (CXS 262-2006), Cheddar (CXS 263-196), Danbo (CXS 264-1966), Edam (CXS 265-1966), Gouda (CXS 266-1966), Havarti (CXS 267- 1966), Samsø (CXS 268-1966), Emmental (CXS 269- 1967), Tilsiter (CXS 270-1968), Saint-Paulin (CXS 271- 1968), Provolone (CXS 272-1968), Cottage Cheese (CXS 273-1968), Cream Cheese (CXS 275-1973) and Cheese (CXS 283-197)

Codex Committee on Pesticide Residues (CCPR50)

- Revision of the Classification of Food and Feed (Type 04: Nuts, seeds and saps Type 05: Herbs and spices)
- Tables on examples of representative commodities for vegetable commodity groups (for inclusion in the Principles and Guidance for the Selection of Representative Commodities for the Extrapolation of MRLs for Pesticides to Commodity Groups) Table 4: Nuts, seeds and saps. Table 5: Herbs and spices
- MRLs for pesticide residues

Codex Committee on Residues of Veterinary Drugs in Foods (CCRVDF24)

- Proposed draft MRLs for amoxicillin (finfish fillet, muscle); ampicillin (finfish fillet, muscle); lufenuron (salmon and trout fillet); monepantel (cattle fat, kidney, liver, muscle)
- Draft RMR for gentian violet

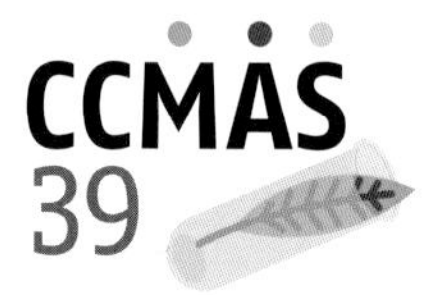

Codex Committee on Methods of Analysis and Sampling (CCMAS39)

- Methods of analysis / performance criteria for provisions in Codex standards
- Editorial corrections